応用航空英語

航空英語能力証明対話試験対策

岩﨑　恵実

鳳文書林出版販売

発刊にあたって

航空英語能力試験の意味

吉田研作
上智大学名誉教授、日本英語検定協会会長、航空英語能力審査会会長

国際線を飛ぶ航空機は様々な国の管制官やパイロットとコミュニケーションの上に成り立っている。以前、インドネシアの航空管制官の話を聞いた。自分たちは一日に 100 か国以上の国の航空機の誘導をしなければならず、それだけの国のパイロットとコミュニケーションすることは大変だ、と言っていたのが印象に残っている。1977 年、スペイン領カナリア諸島のテネリフェ島で起きた KLM と PanAm 機の衝突事故は航空機事故史上最悪の事故と言われている。日航機の御巣鷹山の事故は単一航空機の事故としては最も多くの犠牲者を出した事故だが、テネリフェの事故はジャンボ機 2 機が衝突するという、最悪の事故となった。

この事故は、アメリカ人とオランダ人のパイロットとスペイン人の管制官の間のコミュニケーションがうまく行かなかったことが最も大きな原因だというのが事故調査委員会の結論だった。その後も色々な状況でコミュニケーション上の混乱や誤解が原因で事故は起きてきた。そのため、ICAO（国際民間航空機関）は、2008 年から国際線を飛ぶパイロット及び国際空港の管制官に対して航空英語能力試験を課すようになり、ICAO であらかじめ定めた基準に基づいて加盟各国がそれぞれ独自に実際のテストを作ることになった。日本では、航空輸送技術研究センターの委託を受け、上智大学国際言語情報研究所が中心となり、日本の航空英語能力試験を開発し、現在まで航空局が運用してきた。

航空英語と言っても、本来航空用語としての特殊な英語（ESP）は存在しており、通常の運航上必要な表現としてはすべてのパイロット及び管制官が学んでいる。しかし、問題は、通常の ESP として英語だけでは足らない緊急事態が起きた時のコミュニケーションである。病人がでたり、天候などが急変したり、バードストライクのように、鳥がエンジンに入ってしまい、飛行中にエンジントラブルが起こったり、機内でトラブルが起こったり（ハイジャックなど）した場合、通常の ESP では通用しない。より一般的な英語を使って管制官に事情を説明したり、指示を仰いだりしなければならない。そのような時の英語力を航空能力試験は測っている。なお、英語のネイティブ・スピーカーであれば良いというわけではない。アメリカ人のパイロットとアメリカ人の管制官であれば、普段から使っている口語表現やイディオムを使ってコミュニケーションできるだろうが、パイロットが日本人の場合はネイティブ・スピーカー同士で通じる表現も通じなくなる。つまり、ここでいう航空英語は、慣用表現、口語表現などの文化的特徴を取り除いた、どこの国の人にも通用するいわゆる Plain English でなければならないのである。

本書の著者岩﨑恵実氏は、応用言語学の勉強を土台に、実際に航空英語の評価者としての経験を持っており、そのため本書は具体的で実践的なテキストになっている。

発刊にあたって

森田進治
法政大学理工学部機械工学科航空操縦学専修長　教授
元日本航空 B747-400 機長

パイロットは、国内線、国際線を問わず、管制官と常時連絡を取り合いながら、航空機を安全に誘導し、また誘導されながら目的地に向かって飛行しています。管制の交信は、決して自機のためだけのものではなく、その周辺を飛ぶ国内各社の航空機や、多くの海外からの航空機が、その交信を傍受する事により、周辺状況を正確に把握する事で、安全を担保しています。また、過去に管制官とのコミュニケーションでの齟齬による、事故や重大インシデントが発生したケースも多々報告されていることから、パイロットは、誰にとっても、聞きやすく理解できる英語で話す能力と共に、正確に聞ける能力も必要になります。とりわけ、緊急事態が発生した場合は、自機の状況や管制に対する要求などを、正確に意思疎通をするためには、通常の航空管制用語だけでは対処出来ず、一般的な英語能力が必要となる場面が多々発生します。私自身、国際線のフライトでは、同じ英語でも、米国と英国とでも大きな違いがあったり、地域によっては、非常に聞きにくい発音もあったりと、決して標準化されているとは言えない環境下で苦労した経験をしました。この様に、我々パイロットは、色々な場面で英語という言語を通じて、コミュニケーションが要求されており、自ずと高い英語能力が求められることになります。このような様々な要求の中で、2008 年に「航空英語能力試験」の運用が開始され、現在に至っています。

さて、本書の著者である岩﨑恵実氏は、日本航空で、新人乗員訓練生の英語教官を担当し、当時始まった「航空英語能力試験」に立ち上げから参画されており、航空英語能力証明試験の試験官をされる等、「航空英語能力試験」について非常に深い見識と経験を維持されています。法政大学では、当初、岩﨑氏に「航空英語能力試験」の補習授業として、教育を依頼していましたが、その必要性から、正式に「応用航空英語」の授業を開設し、引き続き岩﨑氏にその教育に携わって頂く様にお願いし、現在に至っています。

本書は、「航空英語能力試験」を受験するに、非常に実践的な内容になっており、初めて受験する人にも、大変わかりやすく解説されています。また、エアライン・パイロットにとっては、定期的に受験する必要があり、久しぶりの受験への準備にも、程よい review になる内容となっています。皆さんによって、本書が有効に活用されることを期待致します。

はしがき

この本は、航空英語能力証明試験を受験する操縦士、航空管制官や学生、訓練生の方々を対象にしています。航空英語能力証明試験は、リスニング試験と対話試験があります。前者では100点満点中7割以上を取得することで、後者ではレベル4以上を取得することで合格とされています。本書は対話試験対策として、これから初めてこの試験を受けようとしている方、現在レベル3でレベル4を目指す方、現在レベル4だけれどもさらに安定してレベル4を取得したいと考えている方、現在レベル4で次回はレベル5以上を目指したい方のために役立つことができれば幸甚です。

ご存知の通り、操縦士ならびに航空管制官は、業務上の通信において、航空管制用語（Air Traffic Control Communications：以下ATCと略す）を用います。国際民間航空機関（International Civil Aviation Organization：以下ICAO）が規定する航空管制用語を使用しないことは、航空無線通信における操縦士と航空管制官のコミュニケーションを阻害する要因となるため使用が義務付けられています。しかし、それと同時に、ATCだけでは対応しきれない事態が生じたときは、一般英語（plain English）を使うことが許されています。“The purpose of phraseologies is to provide clear, concise, unambiguous language to communicate messages of a routine nature. While standardized ICAO phraseologies have been developed to cover many circumstances, mainly routine events but also some predictable emergency or non-routine events, it is important to be clear that it was never intended for phraseologies to fully suffice for all pilot and controller communication need.” ICAO (2004, 7.2.5)

横田（2015）でも同様の指摘をしており、ATCでは、承認・許可・指示は管制用語を用いる一方、情報提供や緊急時における内容など管制用語に明記されていないものは一般英語を使うと述べています。

操縦士および航空管制官の英語コミュニケーション能力不足によって生じる航空機事故を防止するため、ICAOは英語能力証明に関する国際標準を採択しました。これに伴い、2011年3月5日以降、各国の国際線操縦士ならびに航空管制官はICAO航空英語能力証明を保持することが義務付けられました（航空法第33条）。

この試験の本質を鑑みると、普段の業務で使用するATCだけを英語能力証明試験で使用したとしても、レベル4を取得できるものではありません。ATCを基盤としながらも英語を使ったコミュニケーション能力を問うものなので、自分のことばで的確に状況を説明することが求められます。会話はことばのキャッチボールなので、情報を出し惜しむことがあってはいけません。ここのところが普段のATCとは異なる会話力を求められるため、現役の操縦士や航空管制官の方々には戸惑われることがあるのかもしれません。よって、まずはこの試験の特徴は、ATC で対応できると証明することに加えて、一般英語を用いてどんな状況においても英語を用いて対応することができることを証明する必要があると認識することが大切です。

本書の構成

本書は 2 部制で構成されています。第 1 部では、評価項目の捉え方と対話試験における重要な点についてまとめ、状況描写に必要な 3 つのポイントを提示しました。第 2 部では、場面描写の対策として試験に出てくることが想定される場面を提示し、それぞれの場面で重要となる語彙やフレーズをまとめました。また、対話試験で想定される質問や質問の種類についても言及しました。描写例を各場面で紹介していますので、実践的な練習問題として活用してください。最後に間違えやすい用法と対話力上達のコツについてまとめましたのでご参照いただければ幸いです。

出版にあたり、鳳文書林出版販売株式会社社長の青木孝様、イラストを描いてくださった日本航空機操縦士協会運航技術委員の柳井健三様には大変お世話になりました。また、法政大学理工学部機械工学科航空操縦学専修の森田進治教授、山下勝教授には、内容に関するご指導をいただきました。感謝申し上げます。そして、母校上智大学の恩師である吉田研作名誉教授からは、教材作成に関する激励とご指導をいただきこの上なく幸甚です。皆さまの温かいご支援、ご指導があってはじめて本の出版に至ることができました。心よりお礼申し上げます。なお、本文における間違いは全て筆者の責に帰するものです。

目　次

第 1 部

1.1.　持ち時間の有効活用

航空英語能力証明対話試験は 3 部構成で、初めに single picture card、 次に ATC のやり取り、最後に sequence picture card という流れです。Single picture card では 1 枚のカードにある場面の絵が描かれており、その絵を描写することとなります。ATC のやり取りでは、日常の ATC を想定して受験者が操縦士役、試験官が航空管制官役になりフライトで想定される場面でのやりとりを行います。Sequence picture card では、4 コマまたは 6 コマの絵を提示されますので、絵の内容をもとに物語を語るように話します。試験時間は約 15 分ですので、それぞれの場面での持ち時間は約 5 分と考えておきましょう。つまり、15 分の対話試験の中で、自らの英語コミュニケーション能力を証明する必要があるということになります。5 分の中で何をすべきか考えてみましょう。Single picture card を提示された段階で、間髪を入れずに絵の内容を理解しどのような描写をしようか考える必要があります。描写は約 3 分を心掛けましょう。3 分で絵の内容を描写し、必要に応じて自分の経験談を含めるとよいです。読者の方が訓練生ならば、経験談を話すことは難しいかもしれませんが、仮定の話をするのも一つの手です。「もし、私がこのような状況にあるならば…」といった具合に話すことをお勧めします。絵の内容によっては話しやすい、そうでないなどがありますが、どのような絵（状況）であったとしても対応しなければならないことを忘れないでください。

1.2.　評価項目の捉え方と重要な点

対話試験では評価のポイントが、発音（pronunciation）、文構造（structure）、語彙（vocabulary）、流暢さ（fluency）、理解力（comprehension）、対応力（interactions）の 6 つあります（表 1 参照）。航空英語能力証明試験官と評価官をつとめた経験から、この試験では、レベル 4 以上を取得できるかの鍵を握るのは、6 つの評価項目のうち、文構造（structure）、語彙（vocabulary）、対応力（interaction）であると考えています。

＜表 1＞　ICAO Language Proficiency Rating Scale / Operational 4

Pronunciation（発音）	Pronunciation, stress, rhythm, and intonation are influenced by the first language or regional variation but only sometimes interfere with ease of understanding.
Structure（文構造）	Basic grammatical structures and sentence patterns are used creatively and are usually well controlled. Errors may occur, particularly in unusual or unexpected circumstances, but rarely interfere with meaning.
Vocabulary（語彙）	Vocabulary range and accuracy are usually sufficient to communicate effectively on common, concrete, and work-related topics. Can often paraphrase successfully when lacking vocabulary in unusual or unexpected circumstances.

Fluency （流暢さ）	Produces stretches of language at an appropriate tempo. There may be occasional loss of fluency on transition from rehearsed or formulaic speech to spontaneous interaction, but this does not prevent effective communication. Can make limited use of discourse markers or connectors. Fillers are not distracting.
Comprehension （理解力）	Comprehension is mostly accurate on common, concrete, and work-related topics when the accent or variety used is sufficiently intelligible for an international community of users. When the speaker is confronted with a linguistic or situational complication or an unexpected turn of events, comprehension may be slower or require clarification strategies.
Interactions （対応力）	Responses are usually immediate, appropriate, and informative. Initiates and maintains exchanges even when dealing with an unexpected turn of events. Deals adequately with apparent misunderstandings by checking, confirming, or clarifying.

出典：ICAO Language Proficiency Rating Scale (2004) Annex 1, A-8

対応力（interaction）

中でも最も重要なのは対応力（interaction）です。

ホーキンズ（1992）は、航空分野において、「コミュニケーションはヒューマン・ファクター研究の主流」でありながら「言葉によるコミュニケーションにおけるヒューマン・ファクターの認識不足は、新しいことではなく、長年航空業界を悩ましてきたもの」であると述べています。この試験の趣旨である、どのような状況においてもことばを使って説明することができることを証明するためには、コミュニケーション力があることを試験の中で表さなくてはなりません。

わからないことがあれば明確化の要求（clarification, check and confirm）をすることはもちろんですが、情報提供を惜しまないことが重要です。これだけ言えば相手に伝わるだろうと推測しても、情報量が少ないまたは適切でない場合は、情報提供を十分にしているとは言えません。ATC では必要最小限のことを言えば済むかもしれませんが、コミュニケーションが成立するためには、相手にとって過不足ない情報を伝えることが大事です。

例えば、図１を見てみましょう。

・若い女性が国際空港のチェックインカウンターに到着したところである。

・女性は、T シャツを着てサンダルを履いている。

・大きなスーツケースを押しハンドバッグを持っている。

＜図1＞

What is she doing? Please describe. と言われたと仮定します。これに対して A さんと B さんでは、異なる表現をしますが、説明にどのような違いがあるでしょうか。

A：She is a passenger.　She is going to check-in her baggage.　This is an international airport, so she probably is going abroad.

B：This young lady just arrived at the airport.　She's got a big suitcase and a handbag and is going to check-in her baggage.　I don't exactly know where she's going but since this is an international airport, and the fact that she's wearing a T-shirt and sandals, I assume she's going somewhere where it's warm to enjoy her vacation.

正解か不正解かという観点から考えると、A さんも B さんも絵の内容を表現しているのでどちらも正解です。情報量はどうでしょうか。A さんも B さんも 3 文で説明していますが、B さんの方が A さんに比べると、説明が細かいですね。A さんは必要最小限の情報を提供していますが、B さんは形容詞を用いたり、推測した内容を表現したりすることで、この絵に出てくる女性をより細かく表現しています。これがつまり情報を提供するということです。言わなくてもわかるだろうという前提は省かねばなりません。

文構造（structure）

基本的な文構造を構築できること、そしてそれを表現できることが大切です。基本的な文構造は、日本国内で英語教育を受けた方々ならば中学校または高校で学習する範囲の英文法で十分対応できます。例えば、「私は国際線のパイロットです。」と言う場合、例えば I'm a pilot who flies international flights. と 1 文で表現できますが、これを 2 文で表現しても構いません。I'm a pilot. I fly international flights。文章構築が難しい、または関係代名詞の使い方を忘れてしまったという場合であっても、基本文を 2 文用いることで十分対応できます。つまり、文法的に正しい表現をことばに出して説明できるかということが重要なのです。

動詞の時制を一致させることも重要です。今起きていることと想定したならば現在形を、以前にあったことを述べるならば過去形を用いるというように統一させましょう。正しい構文で表現することも重要です。文構造でレベル 4 以上を取得するためには、誤用に注意しましょう。伝えようとした内容とは異なる意味で相手に伝わってしまうのは、誤った文を構築しているからです。

空港の風景を描いた図 2 で考えてみましょう。

・15 分前は大雨に強風と悪天候であった。

・現在は晴天、航空機が離陸している様子がわかる。

<図 2>

A：The weather improved, so the pilot decided to get ready for takeoff.

B：The weather was good, so the pilot decided to get ready for takeoff.

A と B の文は、実際に法政大学理工学部航空操縦学専修の学生を教える中で出た解答例です。A さんも B さんも「しばらくすると天候が回復したので、パイロットは離陸の準備をした。」と表現しようと試みています。どちらが正しいでしょうか。

この絵を表現するためには、A の The weather improved, so the pilot decided to get ready for takeoff（しばらくすると天候が回復した）と言う必要があります。B を表現した学生の意図としては、「しばらくして天候が回復した」と言いたかったと回顧していますが、the weather was good はつまり「天候は良かった（が、今は良くない）」という意味としてとらえられてしまいます。自分としてはうまく表現したつもりでも、実際には相手に正しく伝わっていないことになります。意味に影響を及ぼす誤用は、コミュニケーションが不成立となる原因になりかねませんので、できる限り避けたいところです。

語彙力（vocabulary）

対話試験において語彙力があることは欠かせません。業務上使用する用語は比較的早く容易にことばとして出てきますが、あまり使い慣れていない表現を言おうとすると、ことばに詰まってしまうことがあるかもしれません。言いたいことは頭に浮かんでいるのに語彙力が足りないがゆえに表現できず黙ってしまうのは、この試験では命取りになってしまいます。よって、まずは日常的に語彙力を磨いておくことが大切です。

1.3.　英語能力向上のための 3 つのポイント

ここでは、文構造と語彙力の項目でレベル 4 以上を取得するための 3 つのポイントを挙げます。

ポイント 1：文構造でレベル 4 が取れないまたは弱い（weak level 4）と判定される場合は、単文を用いて短く表現する。

難しい文法を使う必要は全くありません。もちろん言えるのであればそれに越したことはないですが、例えば過去完了、未来完了、仮定法などを使う必要はないですし、重文（二つ以上の単文が同じ資格で結びついた文）や複文（付属する節を持つ文）を使おうと考えなくても対応できます。

以下の文で考えてみましょう。

「風が収まったので、パイロットは滑走路 18 から離陸した。」

A：The wind died down and the pilot took off from runway 18.

B：After the wind died down the pilot took off from runway 18.

C：The wind died down.　The pilot took off from runway 18.

A は重文、B は複文、C は単文です。A は「風が収まったこと」と「パイロットが離陸したこと」を、and を用いることで一つの文にまとめています。一方 B は、after を用いることで、「風が収まったのちパイロットは離陸した。」と表現しています。

一方、C は「風が収まった。」「パイロットは滑走路 18 から離陸した。」と 2 つの文に分けて表現しています。単文とは、主語・動詞の関係が 1 回だけの文を指します。C を言えることができるのであれば十分対応できています。文法が弱い方や文構造でレベル 3 と評価がつく方にお勧めしたい説明方法です。もし読者の方がレベル 5 以上を目指している場合は、単文だけではなく重文や複文を使用しましょう。重要なのは正しい表現で的確に事象を説明できるかという点です。

文構造の問題以外にも、単文を用いて一文の長さを短く区切ることで流暢さ（fluency）を攻略することも可能です。母語でもそうですが、話が長くなるほど、また文が長いほど冗語（fillers）が入ったり、間（pauses）ができたりします。冗語とは、よけいなことばのことを意味しますが、ことばだけでなく「え～」や「あ～」などの音も含みます。評価項目の一つである流暢さでひっかかる場合は、単文を用いて短い文で表現することで、余計なことばや間を回避することができます。

冗語（fillers）が癖になっている場合は、英語で話す時のみならず母語でも同じ兆候が見られるはずなので、普段の話し方を見つめ直し、極力減らす努力をすることが大切です。一度ご自身の話し方を分析してみましょう。

ポイント２：語彙が出てこない時はパラフレーズ（paraphrase）をする。

パラフレーズとは、わかりやすく言いかえることですが、ここで提示するパラフレーズは、Tarone (1977)が提唱したコミュニケーション・ストラテジー（以下、CSs）のことを指します。CSsとは、共通の言語体系を持たない聞き手と話し手が対話の中で、意味を理解し共有することであり（Tarone, 1980）、CSsの中に、パラフレーズ（言い換え）が含まれます。第２言語習得者や外国語学習者が言語を学ぶ過程で、表現したい語彙が浮かんでこないまたはすぐに言えない時のストラテジーとして、パラフレーズを用いることがあると指摘しています。

パラフレーズはapproximation（類語）、 word coinage（造語）とcircumlocution（遠回しの表現）に細分化されます（Tarone, 1978）。このうちの遠回し（circumlocution）とは、語彙そのものを言うのではなく、遠回しに事象を説明することです。例えば、消火器（fire extinguisher）で考えてみると、fire extinguisherと言いたいのに言えない場合、This is used to put out fire、 またはWe use this to extinguish fireという具合に表現します。

パラフレーズを使いこなすためには、普段から練習することをお勧めします。筆者が担当する法政大学理工学部機械工学科航空操縦学専修『応用航空英語』の授業では、パラフレーズの練習を取り入れています。先ほどの消火器の例の他に、悪天候（adverse weather）を別のことばで表現してみるように学生に促し、bad weather such as heavy rain, snow or thunderstormのように言いかえる。これを繰り返すことで、ある事象を説明する中でことばに詰まった際、長い間ができたり立ち止まったりすることがないように意識づけることができます。語彙力を向上させたい方または語彙力が弱い方にお勧めしたい学習法です。Canale and Swain（1980）は、コミュニケーション能力には、４つの要素があると述べています。その４つとは、「文法能力」「談話能力」「社会言語学的能力」「方略的言語能力」のことで、そのうちの「方略的言語能力」とは、会話を継続するために言語学習者がとる手段のことで、コミュニケーション能力の一つであるだけでなく、外国語で話す自信を持つための方略であるとも考えられています。また、Nakatani (2005)は、授業で明示的にCSsを教えることによってコミュニケーションを円滑にとることに対する学生の意識が高まったと報告しています。この先行研究では、発話の内容を相手に確認し、つなぎ言葉や相槌を打つなどの手段を用いることで会話を成立させています。一般的な会話では面と向かって話したり、非言語コミュニケーションを用いたり、つなぎ言葉を用いたりすることで会話を継続させることは日常的なことですが、航空英語ではこれらは通用しません。したがって、航空英語能力証明対話試験の評価項目にある「対応力」でレベル４以上を取得するためには、瞬時に事象を説明できる即答力が必要となります。コミュニケーション能力を鍛えるためには、パラフレーズ練習は欠かせません。

ポイント３：コロケーション（collocation）で語彙やフレーズを覚える。

コロケーションとは、語と語の間における語彙、意味、文法等に関する習慣的な共起関係（堀井、2009）を意味します。ATC の弱点は、無駄を省いた最小限の語彙を用いる特性であるがゆえに、ある事象を文で説明するとなると、ある語と共起関係にある動詞が出てこないなどの問題が生じることと言えます。

言語学では、第二言語または外国語学習者のことを第二言語学習者と呼びます。Selinker（1972）によると、第二言語習得研究の分野では、第二言語学習者は、母語でも目標言語でもない中間言語を使用すると考えられています。中間言語の特徴は、第二言語学習者の母語の影響が転移することです。言語転移には、正の転移（positive transfer）と負の転移（negative transfer）があります。正の転移は、母語と目標言語の言語体系が同じであるため、発音・文法・語用などをそのままスムーズに移行できる一方、負の転移では母語と目標言語の言語体系が異なるがゆえに、母語の言語体系をそのまま目標言語に用いることで誤った用法をしてしまいます。

対話試験における航空英語の場面にあてはめて考えると、負の転移（negative transfer）が起きることがありますが、それはつまり ATC の特徴をそのまま航空英語能力証明の対話試験で用いてしまうからだと考えられます。

例えば、「ダイバートする必要がある」と言う場合に、行き先変更（diversion）を使いたいとします。行き先変更（diversion）は名詞ということに着目してください。

× I need to diversion.

○ I need to divert.

○ I need to make a diversion.

不定詞（to＋動詞）を使わなければならないところで、to の後ろに名詞を置くのは、誤った言い方です。これは実際に筆者が担当する授業を履修していた学生が作った文です。行き先変更（diversion）という単語を使いたかったので文を作ったようですが、文法的に間違った表現になってしまった例です。このようなケースをどうしたら防げるでしょうか？この場合、diversion とセットになれる動詞 make を覚えておくことが誤文を防ぐポイントです。つまり diversion だけではなく make diversion とコロケーションで覚えることが大事です。

コロケーションは動詞だけでなく形容詞と名詞のセットを覚えるのにも役立ちます。例えば、「強風」は英語で strong wind と言いますが、heavy wind とは言いません。一方、「豪雨」は英語で heavy rain と言いますが、strong rain とは言わないので不自然な言い方と捉えられてしまいます。意味の理解に影響を与えることはありませんが、英語学習をするならば正しい表現で覚えることを心がけましょう。

第 2 部

2.1.　運航場面の 6 つの局面（フェーズ）

航空英語能力証明試験では、主に 6 つの運航場面のフェーズ（表 2 参照）が問題に出されます。絵には、日常業務の場面に加えて、非日常の場面（non-normal situation）が描かれています。

＜表 2＞　6 つの運航場面のフェーズ

Phase 1	Before Takeoff (At the airport or ground)
Phase 2	During Takeoff
Phase 3	During Cruising
Phase 4	Before Landing
Phase 5	During Landing
Phase 6	After Landing

対話試験では、表 2 に示した 6 つの運航場面のフェーズの中で、遭遇するまたは遭遇する可能性のある出来事が single picture card に提示されます。そこで、第 2 部では、picture card を場面別に見ていきます。場面を提示しますので、それを見て、自分だったらどのように描写するか考えてみましょう。そして、おさえておきたい単語と各場面で想定される質問を使って説明できるか確認してみましょう。

2.2.　質問のタイプ

まず初めに、質問の種類についておさえておきましょう。ここでは、代表的な質問のタイプを 4 つ提示します。

1. Closed Question

閉じた質問とは、答えが 1 つしかない質問のことを指します。

例）Are you a pilot？　Yes, I am.

2. Open Question

開いた質問とは、質問に対する答えが一通りではありません。

例）What did you do yesterday？　I stayed home and took a rest.

3. Display Question

事実質問とは、聞き手が答えを知った上で敢えてする質問のことを指します。

例）How is the weather at NRT today?　It's sunny and no chance of rain.

4. Referential Question

指示質問とは、聞き手が答えを知らずにたずねる質問のことを指します。

例）Why are you in a hurry?　I have to catch a train to get to work.

対話試験では、上記の質問の種類のうち、2 と 4 が頻繁に用いられると考えて間違いありません。1 と 3 を使う場合もありますが、多くの場合 open question と referential question が用いられます。なぜでしょうか？一言でいえば、話し手の説明能力を問う問題を出したいからです。よって、yes, no で答えが完結する質問はほとんどでません。質問形式は 5W1H（who, what, when, where, why, how）が基本となります。

2.3.　場面別 picture card　描写対策

ここからは状況描写の実践練習です。Exercise A と B の流れで進めてみましょう。

【Exercise A】

1. 図を見て状況を描写する。注 1)
2. インタビューアーになったつもりで質問を、受験者になったつもりで質問に対する答えを考えてみる。
3. ペアで練習している場合は、それぞれの質問と答えを比べてみる。

【Exercise B】

1. 描写例（description example）を確認する。注 2)
2. 描写例を声に出して練習する。
3. インタビューアーから想定される質問（possible questions）に答えてみる。
4. 解答例（model answers）、ポイントとなる語彙（key vocabulary）、表現（useful expressions）を確認する。

注 1）実際の対話試験では、イラストに対する日本語の説明文は入りません。

注 2）Description Example はあくまでも一例ですので、このようでなければならないというものではありません。

≪描写時のポイント≫

・描写をする時には自分の声を録音し、後で聞き返してみる。
・描写時間を計ってみる。（描写時間の目安は 3 分と考えましょう。）
・描写すべきことは過不足なく表現する。
・描写しにくいまたは表現できない言葉をメモし、調べてみる。

Situation（想定場面）

Weather problems (heavy snow, thunderstorm)

Passenger problems (injured, sick, unruly)

Natural disaster (earthquakes, volcanic eruption)

Pre-flight problems (birds and other animals, pre-flight inspection)

Engine problems (engine failure)

Weather Problems

場面 1：Heavy Snow

・航空機の翼と胴体に雪が積もっている。

・駐機場の周りも雪が積もっている。

・除雪車が航空機およびエプロンの除雪作業を行っている。

<図 3>

【Exercise A】

1. Describe the picture.
2. Think of three possible questions and three possible answers to the questions.
3. If you are working in pairs, compare each question and answer with your partner.

Q1. ______________________________

Q2. ______________________________

Q3. ______________________________

A1. ______________________________

A2. ______________________________

A3. ______________________________

【Exercise B】

Description Example

I can see an aircraft at the apron. Snow has accumulated on the plane. We need to remove snow from the body and wings because it is too heavy and it's dangerous to have snow on the airplane. To remove snow, we spray de-icing fluid by using a snow removal truck. It probably takes about 30 minutes to remove the snow completely. At the same time, snow has to be removed from the runway and taxiways as well. It is necessary because snow will eventually turn into ice and the roads will be slippery. Usually, we don't expect to have heavy snow around the Kanto area but it is common in the northern parts of Japan.

Possible Questions

1. What is de-icing?
2. What is anti-icing?
3. What is the difference between de-icing and anti-icing?
4. How do you remove snow from an airplane?
5. What preparations do you need to make for winter operations?

Model Answers

1. De-icing means to remove snow from an airplane.
2. Anti-icing means to prevent snow from accumulating on an airplane.
3. De-icing is to remove snow from an airplane whereas anti-icing is to prevent snow from accumulating on an airplane.
4. We spray de-icing fluid on it.
5. There are lots of preparations needed for winter operations. For example, getting information about the snow condition at the arrival airport, getting ready for de-icing and anti-icing and preparing for diversion.

≪Key vocabulary≫

Verb

・accumulate, pile up（積もる）
・remove（取り除く）
・slip（滑る）
・spray（〜をかける）

Noun

・anti-icing（防氷）
・de-icing（除氷）

・de-icing fluid（除氷剤）

・snow removal（除雪作業）

Adjective

・slippery（滑りやすい）

≪Useful Expressions≫

1. remove A from B

2. spray 目的語 by～ing

3. ～means…：～とは…を意味する

4. to prevent～from 動詞 ing：～が（動詞）するのを防ぐ

5. whereas：一方

場面 2：Thunderstorm

・飛行場に 4、5 機の航空機が駐機している。

・天候は大雨に、強風

・飛行場近くの空模様は積乱雲が見えている。

【Exercise A】

1. Describe the picture.
2. Think of three possible questions and three possible answers to the questions.
3. If you are working in pairs, compare each question and answer with your partner.

Q1. ______________________________

Q2. ______________________________

Q3. ______________________________

A1. ______________________________

A2. ______________________________

A3. ______________________________

【Exercise B】

Description Example

I can see some airplanes parked at the apron. The weather seems very bad with heavy rain and strong wind. We can tell the strength of the wind by looking at the windsock. I can also see big clouds, possibly cumulonimbus (CB). CB is very dangerous to aircrafts and we must try to avoid this cloud. Heavy rain and strong wind are sometimes accompanied by lightening. If lightening hits an aircraft, the aircraft system will be affected so we must avoid lightening as well. For visual flight rules (VFR) flights, we would never fly in this weather condition.

Possible Questions

1. Why is a thunderstorm dangerous to aircrafts?
2. How do you get information about a thunderstorm?
3. Is it possible to fly in thunderstorm weather?
4. What would you do if the weather at the destination airport is bad?
5. If you have to divert to another airport, what would you tell your passengers?

Model Answers

1. It is difficult to fly in a thunderstorm. It's very windy and lightening might affect the aircraft system.
2. We have lots of sources of information, such as ATC, company radio and PIREP (pilot report).
3. It depends on the condition but the best thing is to avoid thunderstorm areas because there will

be heavy rain, strong wind and possibly lightening and microbursts.

4. If it's very bad, I'll request a diversion to an alternate airport.
5. I will tell the passengers, due to bad weather at the destination airport, we need to divert to ～ airport.　We apologize for the inconvenience. Thank you for your patience and understanding.

≪Key vocabulary≫

Verb

・flash（稲妻が走る、光る）

・strike（うつ）

Noun

・heavy rain（強雨）

・strong wind（強風）

・windsock（風見用円錐筒、吹き流し）

・lightening（稲妻）

・microburst（瞬間突風）

Adverb

・possibly（ことによると、おそらく）

≪Useful Expressions≫

1. 主語 seems～：主語が～のようにみえる
2. tell A by 動詞 ing：A を（動詞）することによってわかる
3. it depends on～：～次第
4. request～：～を要請する
5. due to～：～によって

Passenger Problems

場面 3：Injured Passenger

・CA が食事を提供している。

・カートからお湯がこぼれ、一人の乗客にかかってしまう。

・お客様は服が濡れてしまい、かつ熱いお湯に大変びっくりしている。

<図５>

【Exercise A】

1. Describe the picture.
2. Think of three possible questions and three possible answers to the questions.
3. If you are working in pairs, compare each question and answer with your partner.

Q1. ______________________________

Q2. ______________________________

Q3. ______________________________

A1. ______________________________

A2. ______________________________

A3. ______________________________

【Exercise B】

Description Example

This was during cruising.　The seatbelt sign was off and the cabin attendants were serving food to the passengers.　However, there was a sudden jolt or light turbulence and unfortunately hot water that was on the cart splashed onto one of the passengers who was sitting in an aisle seat. It was hot water so the passenger was shocked.　His clothes got wet and he started to get upset. He got angry at the cabin attendant and complained how hot it was.　The cabin attendant brought towels and ice for the passenger and was very apologetic.

Possible Questions

1. What do you think has happened to the passenger?
2. If a passenger gets hurt during flight, what action do you take?
3. If there is no doctor on board, what would you do?
4. What things are included in the first-aid kit?
5. What kind of injuries are possible in an airplane?

Model Answers

1. Hot water spilled on the passenger while a cabin attendant was serving drinks.
2. First, I need to get detailed information about the passenger's condition from a cabin attendant. Depending on the situation, we have to ask for a doctor's help by making a doctor call.
3. I will ask for the company doctor's advice.
4. I don't exactly know but most likely bandages, plasters, disinfectant, and aspirin.
5. There are many possibilities such as hitting one's head, bumping one's legs, getting cut, burning oneself with hot water or drinks to name a few.

≪Key vocabulary≫

Verb

・hurt（けがをする、痛める、痛む）

例）He hurt his leg.（彼は足を痛めた。）　His leg hurts.（彼の足が痛む。）

・spill（こぼす）

・bump（ぶつかる）

・主語 splash onto～（主語が～にかかる）

Noun

・bandage（絆創膏）

・plaster（ギプス、包帯）

・disinfectant（消毒剤）

・burn（やけど）
・bruise（あざ）
・aspirin（鎮痛剤）

Adjective
・painful（痛い）

≪Useful Expressions≫

1. get 形容詞：形容詞の状態になる
2. ask for～：～をたずねる
3. make a doctor call：ドクターコール（医者がいないかの声掛け）をする
4. I don’t exactly know but～：詳しくは知らないが～
5. to name a few：いくつか例を挙げると

場面 4：Sick Passenger

・あるお客様の調子が悪そうである。
・頭痛と腹痛の両方にみまわれて、吐き気もする様子。
・CA がお客様の状況をパイロットに報告している。

＜図６＞

【Exercise A】

1. Describe the picture.
2. Think of three possible questions and three possible answers to the questions.
3. If you are working in pairs, compare each question and answer with your partner.

Q1. ______________________________

Q2. ______________________________

Q3. ______________________________

A1. ______________________________

A2. ______________________________

A3. ______________________________

【Exercise B】

Description Example

One of the passengers looks very pale. He is sweating a lot and seems to have both a headache and stomachache. He looks like he wants to throw up any moment. The chief cabin attendant is reporting the passenger's situation to the pilots in the cockpit by using the intercom. The captain made a doctor call but there was no doctor on board. With the information they get from the cabin, the captain and the co-pilot are discussing whether to go on to the destination airport or divert to another airport so that the sick passenger can be taken to a hospital as soon as possible.

Possible Questions

1. What do you think is happening with this passenger?
2. What would you do if a passenger gets sick during flight?
3. Do you have any advice to a passenger who is suffering from motion sickness?
4. What would you do if you get sick during your flight?
5. In what cases, do you have to make an emergency landing?

Model Answers

1. He is sweating a lot and looks pale. Perhaps, he is suffering from motion sickness.
2. If I get a report from a cabin attendant that one of the passengers is sick, first I will confirm the passenger's condition. If the situation is mild, I'll ask a cabin attendant to take care of the passenger and keep me informed. If the situation is bad, I might need to make an emergency

landing.
3. Try not to read small letters, take enough liquid.
4. I'll tell the captain (or co-pilot) about it and take a little rest.
5. When a passenger has a heart attack or loses consciousness, he or she needs to be taken to a hospital immediately.

≪Key Vocabulary≫
Noun
・motion sickness（乗り物酔い）
・heart attack（心臓発作）
Adjective
・sweating（汗をかいている）
・pale（青白い、顔色が悪い）

≪Useful Expressions≫
1. discuss whether to A or B：A または B について話し合う
2. ask 〜 to 動詞：〜に（動詞）してもらうよう頼む
＊keep me informed：常に情報・状況を報告してもらう
3. make an emergency landing：緊急着陸をする
4. have a heart attack：心臓発作が起きる
5. lose consciousness：意識を失う

場面 5：Unruly Passenger
・トイレ待ちのお客様が 2、3 名いる。
・トイレのドアのサインは使用中
・トイレの中では、たばこを吸おうとしているお客様がいる。
・煙探知機が天井に設置されている。

＜図 7＞

【Exercise A】

1. Describe the picture.
2. Think of three possible questions and three possible answers to the questions.
3. If you are working in pairs, compare each question and answer with your partner.

Q1. ______________________________

Q2. ______________________________

Q3. ______________________________

A1. ______________________________

A2. ______________________________

A3. ______________________________

【Exercise B】

Description Example

This was during cruising and perhaps after a meal because most of the passengers look relaxed with some beverage in front of them. Two or three passengers are waiting outside a lavatory. The sign at the lavatory door says occupied. Apparently, one passenger is going to start smoking in the lavatory. I can tell this because he has a lighter and tobacco in his hand. I don't know what's going to happen after this because there is a smoke detector in all the lavatories and if he starts smoking, the alarm will definitely ring. Smoking is absolutely prohibited in an airplane therefore, if he starts smoking, he will be in big trouble.

Possible Questions

1.What is the problem with this passenger?

2. If you get a report from a cabin attendant that there is a passenger who is smoking in the lavatory, what would you do?

3. Why is smoking prohibited in an airplane?

4. What other unruly behavior can you think of?

5. How do you handle unruly passengers?

Model Answers

1. He is smoking in the lavatory which is against the law.

2. I will tell the cabin attendant to give warning to this passenger. He has to know that smoking is prohibited.

3. Smoking in an airplane is dangerous because it might cause a fire.

4. Listening to loud music, becoming drunk and causing trouble to other passengers, not putting the reclining chair back to its original position during meals or takeoff and landing and not fastening the seatbelt when the seatbelt sign is on.

5. First of all, I tell a cabin attendant to report the situation. If a passenger doesn't listen to the warning and keeps causing trouble, then the passenger will have to be restrained according to aviation law (rules).

≪Key vocabulary≫

Verb

・handle（対応する）

・warn（忠告する）

・restrain（拘束する）

Noun

・reclining chair（リクライニングチェア）

Adjective

・unruly, disorderly（規則にしたがわない、手に負えない）

・unhealthy（健康に悪い）

Adverb

・absolutely（絶対に）

・besides（その上）

≪Useful expressions≫

1. be in big trouble：大きな問題となる
2. against the law：法律違反
3. give warning to～：～に対して警告する
4. cause troubles：問題を起こす
5. put ～ back to…：～を…の状態に戻す

Natural Disasters

場面 6：earthquakes

・誘導路（taxiway）に航空機が 2、3 機止まっている。

・1 機の航空機に着目し、コックピットにいるパイロットが地震の情報を入手した。

・空港の上（空）では、着陸待ちの 3 機の航空機が空中待機（holding）している。

＜図8＞

【Exercise A】

1. Describe the picture.
2. Think of three possible questions and three possible answers to the questions.
3. If you are working in pairs, compare each question and answer with your partner.

Q1. ____________________

Q2. ____________________

Q3. ____________________

A1. ____________________

A2. ____________________

A3. ____________________

【Exercise B】

Description Example

Several aircrafts are stopped at the taxiway. There seems to have been an earthquake and the pilots in the cockpit are receiving information concerning the earthquake. Their faces look serious. Perhaps it might have been a rather big earthquake. At the same time, I can see 3 airplanes holding over the airport. They are waiting for the ATC clearance to land at the airport. However, because there was an earthquake, it might take some time before these airplanes can make a landing.

Possible Questions

1. How do you get information about earthquakes during flight?
2. What happens to an airport when there is a big earthquake?
3. Do earthquakes affect your flight? If so, how?
4. If there is a crack on the runway, what should be done?
5. If you cannot land at the destination airport due to a big earthquake, what options do you have?

Model Answers

1. We get information about earthquakes from ATC and the company.
2. The airport might be closed for a while because it needs a thorough inspection.
3. Depending on how bad the quake was, it certainly does. We might not be able to land at the destination airport and instead, we have to divert to another airport.

4. The JCAB (Japan Civil Aviation Bureau) will come to the airport and check the runway and taxiway conditions.　If there is a crack on the runway, no aircrafts can land there.　After they do a runway sweep, they need to seal cracks.
5. We will have to divert to another airport.　The alternate airport is usually the closest airport from the original airport.

≪Key vocabulary≫

Verb
・receive（受け取る）
・divert（迂回する）
・restrain（拘束する）
・seal（補修のため埋める）

Noun
・inspection（調査）
・runway sweep（滑走路をきれいにする）
・alternate airport（代替飛行場）

Adjective
・thorough（徹底的な）

Adverb
・rather（いくぶん、やや）

≪Useful expressions≫

1. there seems to have been～：～があったようだ
2. it might have been～：～であったかもしれない
3. at the same time：同時に
4. it might take some time before～：～する前に少し時間がかかるかもしれない
5. divert to～：～にダイバートする（代替飛行場に行く）

場面 7：volcanic eruptions

・桜島が噴火（小規模）し、火山灰があがっている。
・周辺を飛行中の航空機が噴火の情報を ATC から入手している。
・航空機が旋回し、航路を変更している。

＜図9＞

【Exercise A】

1. Describe the picture.
2. Think of three possible questions and three possible answers to the questions.
3. If you are working in pairs, compare each question and answer with your partner.

Q1. ______________________________

Q2. ______________________________

Q3. ______________________________

A1. ______________________________

A2. ______________________________

A3. ______________________________

【Exercise B】

Description Example

Several jet aircrafts are cruising.　Far away, I can see a mountain.　It looks like it's not just a mountain but a volcano.　Perhaps, mount Sakurajima.　It is an active volcano.　Pilots are trying to get as much information as possible from the ATC and the airline company.　I can see volcanic ash clouds surrounding the mountain.　Volcanic ash clouds are extremely dangerous to aircrafts.　We certainly don't want any small particles entering the engine.　The best thing for the pilots to do is to avoid flying around or near the mountain.

Possible Questions

1. How do you get information about volcanic eruptions during flight?
2. What other sources of information do you rely on to know about volcanos and volcanic ashes?
3. How does a volcanic eruption affect your flight?
4. Why is a volcanic eruption dangerous to aircrafts?
5. Is there anything else you need to be careful of concerning mountains?

Model Answers

1. We get information from ATC and the airline company.
2. We can also refer to VAAC (Volcanic Ash Advisory Center) and the Japan Meteorological Agency to get the latest information as well as model cases from the past.
3. If we get information about a volcanic eruption, then we have to think of changing the flight course. The important thing is to avoid the eruption.
4. Eruptions are very dangerous because volcanic ashes include minute particles that cannot be seen and if they enter the engines, the engines will be damaged.
5. Around the mountains, there might be sudden windshear.

≪Key vocabulary≫

Verb

・erupt（噴火する）
・avoid（避ける）
・cause～（～の原因となる）
・include～（～を含める）
・affect～（～に影響を与える）

Noun

・volcano（火山）
・volcanic eruption（火山の噴火）

・volcanic ashes（火山灰）
・particles（微粒子）

Adjective

・minute（微小な）

≪Useful expressions≫

1. it's not A but B：A ではなく B
2. the best thing for ～to do is…：～にとって最良の方法は…
3. ～cannot be seen：～は目に見えない
4. ～will be damaged：～がダメージを受ける
5. it could cause～：～を引き起こすかもしれない

Problems with Aircrafts & Runways

場面 8：Birds on runways

・滑走路で鳥がえさをついばんでいる。
・パイロットが ATC に bird sweep を頼んでいる。
・グランドスタッフが車でかけつけ鳥を追い払おうとする。

＜図 10＞

【Exercise A】

1. Describe the picture.
2. Think of three possible questions and three possible answers to the questions.
3. If you are working in pairs, compare each question and answer with your partner.

Q1. ____________________

Q2. ____________________

Q3. ____________________

A1. ____________________

A2. ____________________

A3. ____________________

【Exercise B】

Description Example

The aircraft is heading toward the runway. Unfortunately, I can see some birds pecking on the runway. I don't think this aircraft can take off immediately because pilots want to avoid a bird strike especially just before takeoff. First, the pilots have to ask the ATC for a bird sweep. A bird sweep is to sweep birds from a runway by using an air gun. Most of the time, the birds will fly away because they get scared by a big sound. After the bird sweep is done, this aircraft can take off from this runway.

Possible Questions

1. What is a bird strike?
2. What is a bird sweep?
3. Why are birds dangerous to aircrafts?
4. If you see lots of birds on runway before takeoff, what do you do?
5. Are there any other animals you have to be careful of?

Model Answers

1. A bird strike means a bird hits the airplane. A bird gets sucked into the engine. This is usually the case after a bird strike. If it happens, the engine might catch fire.
2. A bird sweep means to remove birds from runways or taxiways to make a clear space by using air guns. Birds get scared of big sounds.

3. They could cause engine fire.
4. I will ask the ATC controller for a bird sweep.
5. Sometimes, deer, foxes, and stray dogs can be seen at certain airports.

≪Key vocabulary≫

Verb

・hit（ぶつかる）
・get sucked into～（～に吸引される）
・ask for～（～を頼む）

Noun

・bird strike：（バードストライク）
・bird sweep：（バードスイープ）
・stray dogs：（野良犬）
・air gun：（エアガン）

Adjectives

・certain（ある…）

≪Useful expressions≫

1. I don't think～：私は～だとは思わない
2. ～is to sweep 目的語 from 場所 by 動詞 ing : ～は目的語を動詞 ing することで、ある場所から追い払う
3. most of the time：ほとんどの場合
4. 主語 get scared by～：主語が～を怖がる
5. ～is usually the case：～はよくあること

場面 9：Pre-flight inspections

・航空機（Cessna172）がエプロンに駐機している。
・パイロット（女性）が航空機の外に出て、機体の状態を目視でチェックしている。
・パイロットの手にはチェックリストがある。
・傍で教官が様子を見ている。

<図 11>

【Exercise A】

1. Describe the picture.
2. Think of three possible questions and three possible answers to the questions.
3. If you are working in pairs, compare each question and answer with your partner.

Q1. ______________________________

Q2. ______________________________

Q3. ______________________________

A1. ______________________________

A2. ______________________________

A3. ______________________________

【Exercise B】

Description Example

An aircraft is parked at the apron.　This is a small aircraft, perhaps, Cessna 172.　The pilot of this aircraft is doing a visual check.　She is checking for any signs of defects or scratches.　If she finds something, she has to report it and have an engineer check the problem. A pre-flight inspection is an important process for pilots.　They have to be responsible for the aircrafts and make sure they are in good condition.

Possible Questions

1. What is the pilot doing?
2. What does a pre-flight inspection mean?
3. What do you do for the pre-flight inspection?
4. Why is it important for a pilot to do a pre-flight inspection?
5. If you find some problems, what do you do?

Model Answers

1. The pilot is checking the condition of the airplane she's going to fly today.
2. Pre means before and inspection means to check something.　A pre-flight inspection means to check the condition of an aircraft before flying.
3. We do an outside check to see whether there are any dents or cracks on the body of an aircraft. We also check the tires and engine to make sure they are in good condition.
4. Safety is more important than anything else, therefore it is vital for pilots to check the condition of an aircraft before flight.
5. We have to report it to our company immediately and have the maintenance staff check the status.

≪Key vocabulary≫

Verb

・report～：（～を報告する）

Noun

・inspection：（点検）
・dents：（へっこみ）
・cracks：（ひび）
・maintenance staff：（整備士）
・status：（状態、状況）

Adjective

・vital：（重要な）

Conjunction

・therefore：（したがって）

≪Useful expressions≫

1. be responsible for～：～の責任を持つ

2. the condition of～：～の状態

3. ～is more important than anything else：他の何よりも～が大切である

4. be vital for～：～にとって重要である

5. have A check～：A が～をチェックする

＊have は使役動詞

場面 10：Engine Failure

・航空機のエンジンから火が出ている。

・パイロットが緊急着陸を要請している。

・ATC がそれに対応し、空港では緊急配備がされている。

<図 12>

【Exercise A】

1. Describe the picture.
2. Think of three possible questions and three possible answers to the questions.
3. If you are working in pairs, compare each question and answer with your partner.

Q1. ______________________________

Q2. ______________________________

Q3. ______________________________

A1. ______________________________

A2. ______________________________

A3. ______________________________

【Exercise B】

Description Example

This is just before landing. I can see fire from one of the engines of the aircraft. I don't exactly know the reason for this fire, but it is a dangerous situation. The pilots are requesting an emergency landing. ATC is responding to the situation. Both fire engines and ambulances will be standing by near the runway. After landing, all the passengers and crew members need to get out of the aircraft immediately.

Possible Questions

1. What do you think has happened to this aircraft?
2. What is an engine failure?
3. Why does engine failure occur?
4. If you have an engine failure during your flight, what do you do?
5. What is your role during emergency evacuation?

Model Answers

1. I think something went wrong with the engine because I can see fire coming out from one of the engines.
2. Engine failure means there is a malfunction with the engine. If the situation is bad, it could catch fire.
3. There are many reasons for an engine failure to occur. For example, a bird strikes the engine,

electric system goes wrong, volcanic ashes enter the engine, to name a few.

4. I need to make an emergency landing immediately. If the passengers are on board, I have to let everybody know about the situation and have them get ready for an emergency evacuation after landing.
5. The pilots have to shut down the engine, ask for immediate help on the ground and make sure all the passengers get out of the aircraft safely.

≪Key vocabulary≫

Verb

・go wrong：(問題が発生する)
・come out from～:（～から出てくる)
・～occur：(～が生じる)
・shut down～：(～を停止する)
・get out of～：(～から脱出する)

Noun

・malfunction：(不調、機能しない)

Adverb

・immediately：(すぐに)

≪Useful expressions≫

1. just before～：～の直前
2.～will be standing by：～が待機している
3. get out of～：～から出る
4. something goes wrong with～：～がおかしくなる、～に不具合が生じる
5. make sure～：～かどうか確かめる

2.4. 模擬練習問題

Describe the picture and then answer the questions.

<図13>

【Questions】

1. If you are the pilot in this picture, what concerns will you have for today's flight?
2. What does hard landing mean?
3. What does windshear mean?

【Your answers】

A1. ______________________________

A2. ______________________________

A3. ______________________________

模擬練習問題の描写例と解答例

Description Example

This is a dispatch room. I can see several dispatchers and pilots who are talking about their flights for the day. The dispatcher and the pilot at desk A are looking closely at the monitor. The pilot is scheduled to have a flight to Naha airport but apparently, a typhoon is coming toward Okinawa. The dispatcher is explaining the flight plan such as altitude, amount of fuel, flight course and flight time in detail. Most likely, the weather at the destination airport will be bad so the pilot has to be aware of the latest weather report.

Model Answers

1. We can expect heavy rain and very strong wind at the destination airport and so we might have a hard landing.
2. Hard landing means an airplane cannot land smoothly and there might be strong impact at the time of landing. It happens when there is a crosswind or windshear.
3. Windshear means sudden change in wind direction and velocity. There is also CAT (clear-air turbulence) which occurs out of nowhere when there are no visible clouds around.

2.5.　間違えやすい用法＆対話力上達のコツ

１．Almost（ほとんど）副詞

「ほとんどの」と言う場合は、almost all (of) the または most of the となる。

例）ほぼ全員の乗客が搭乗している。

× Almost passengers are on board.

○ Almost all the passengers are on board.

○ Most of the passengers are on board.

２．複数形の主語に合わせた動詞の用い方

英語の動詞は、主語が単数形か複数形かで活用形が異なります。本当は複数形で話すべきところを単数形扱いにしているケースが多々見られます。

例）3 機の航空機が飛行場の上を空中待機していた。

× Three aircrafts was holding over the airport.

○ Three aircrafts were holding over the airport.

＊日本語に複数形の概念がないため、英語で話す時にも単数形の動詞を用いてしまうのは、負の転移（negative transfer）の例と言えます。

３．動詞が抜ける

英文で話す時に、動詞が抜けているケースがよくあります。航空管制通信（ATC）では簡略化したことばで話すことになっているため、一般英語で話す場合と乖離していますが、英文を作る際には、動詞は重要です。

例）3 機の航空機が飛行場の上を空中待機している。

× Three aircrafts holding over the airport.

○ Three aircrafts are holding over the airport.

＊are がなくても意味は通じますが、正しい英文法ではありませんので、be 動詞＋動詞 ing と用いましょう。

４．主語に誰/何を持ってくるかによって、構文が変化します。

例えば、「航空管制官がパイロットに飛行場の上空で待機するよう指示した」という文を言うならば、次の①や②の文で表現できます。

① <u>The ATC</u> controller told the pilot to hold over the airport.

② <u>The pilot</u> was told by the ATC controller to hold over the airport.

①と②の違いは、能動態か受動態かという点です。主語が異なることで、能動態の文になるか受動態になるかが変わります。上記の文では①が能動態、②が受動態です。

5．自動詞と他動詞

動詞には自動詞と他動詞があるので、注意が必要です。同じ語であっても自動詞にも他動詞にもなり得ます。自動詞は動詞の直後に目的語をとらない一方、他動詞は目的語を必要とします。何を主語に持ってくるかで使い方が変化しますので注意しましょう。

例）飛行機が安全に着陸した。

The airplane landed safely. （自動詞）

I landed the airplane safely. （他動詞）

6．For example と言ったのならば、例を挙げたのち理由や内容を述べましょう。例だけを挙げても物足りなさが残ります。例に対する理由付けをすることで説得力が増すとともに対応力があると示すことができます。

7．Sequence picture card（4 コマまたは 6 コマの絵）を用いて物語を作る時に、インタビューアーが "This happened last week"と指定してくる場合があります。この時は、過去に起きたことと想定して過去形を用いましょう。現在形と過去形を交えて話さないことがコツです。時制を統一できているかをみています。

8．3 分話すことは短いようで長く、長いようで短いと言えます。試験を受ける方の中には、「日本語でさえ 3 分も話すのは難しい」とおっしゃる方もいますが、その時は、次の 3 つのポイントを覚えておいてください。

① 絵を見て、見たままの事実を述べる。

② 想像力をふくらませて絵を脚色する。

③ 実際に体験したことを話す。または、似通った体験があるならば体験談を交える。

もし体験したことがないならば、仮定の話をしてみましょう。

例）「私自身は go around の経験はないが、もし同じ状況にあったならば、確実に安全着陸するため go around するでしょう。」

I myself do not have an experience of making a go around, but if I were in this situation, I would make a go around to make sure I can land safely.

・if I were in this situation, I would～：もし私がこの状況にあったならば、～するだろう

引用・参考文献

Canale, M., & Swain, M. (1980). Theoretical bases of communicative approaches to second language teaching and testing. *Applied Linguistics*, 1, 1-47.

International Civil Aviation Organization. (2004). Manual on the Implementation of ICAO Language Proficiency Requirements. Doc9835.

Internationl Civil Aviation Organization. (2010). Manual on the Implementation of ICAO Language Proficiency Requirements. Doc9835, AN/453.

MacIntosh, C., & Francis, B., & Poole, R. (Eds). (2009). *Oxford Collocations Dictionary*. Oxford: Oxford University Press.

Nakatani, Y. (2005). The effects of awareness-raising on oral communication strategy use. *The Modern Language Journal*, 89, 76-92.

Selinker, L. (1972). Interlanguage. IRAL, 10 (2), 209-231.

Takatsuka, S. (1999). Teaching paraphrase as a communication strategy: a critical review and proposal. *Annual Review of English Language Education in Japan* (ARELE), vol. 10, 21-30. The Japan Society of English Language Education.

Tarone, E. (1977). Conscious communication strategies in interlanguage: a progress report. Brown, H., C. Yorio, & R. Crymes (Eds.), *Teaching and Learning English as a Second Language*, TESOL. 194-203. Washington, D.C.

Tarone, E. (1978). Conscious Communication Strategies in Interlanguage: A Progress Report, in H.D. Brown, C. Yorio & R. Crymes (Eds.), On TESOL "77: *Teaching and Learning English as a Second Language*, Washington, D.C.: TESOL.194-203.

Tarone, E. (1980). Communication Strategies, Foreigner Talk, and Repair in Interlanguage. *Language Learning*, 30, 417-431.

操縦士等に対する航空英語能力証明試験モデル開発調査研究委員会 (2006).『操縦士等に対する航空英語能力証明試験モデル開発調査研究報告書』 航空輸送技術研究センター.

フランク・H. ホーキンズ (1992).『ヒューマン・ファクター－航空の分野を中心として』成山堂書店.

鳳文書林出版 (2020).『航空法』

堀正広 (2009).『英語コロケーション研究入門』研究社.

横田友宏 (2015).『エアラインパイロットのための ATC』鳳文書林出版.

≪MEMO≫

≪MEMO≫

≪MEMO≫

≪MEMO≫

≪MEMO≫

≪MEMO≫

【筆者プロフィール】

岩﨑　恵実（いわさき　えみ）

上智大学外国語学部英語学科卒業。
上智大学大学院外国語学研究科言語学専攻博士前期課程修了。
専門分野は応用言語学および航空英語教育。
元日本航空株式会社（当時：株式会社日本航空インターナショナル）運航本部運航乗員訓練部基礎教官室英語教官。航空会社在職中、航空英語能力証明試験試験官および国土交通省認定能力判定員を担当する。航空会社退職後、千葉県私立高校での教員を経て、現在秀明大学観光ビジネス学部専任講師、および法政大学理工学部航空操縦学専修非常勤講師。

【イラストレーター】

柳井　健三（やない　けんぞう）

川崎重工業入社、元パイロット。
小型機、大型機、練習機（多機種）に 44 年間、計 10,650 時間乗務する。
飛行機：YS11、P2V7、P2J、P3C、E2C、C130、C1、C1-STOL など。
回転翼機：Bell47、H369、V107、CH47、EC120 など。

発行　令和３年10月29日　　　印刷　シナノ印刷

応用航空英語

航空英語能力証明対話試験対策

岩﨑　恵実著

発行　鳳文書林出版販売

〒105-0004　東京都港区新橋３－７－３

Tel 03-3591-0909　Fax 03-3591-0709　E mail info@hobun.co.jp

ISBN978-4-89279-466-7 C3032　¥1800E　　定価 1,980 円（本体 1,800 円＋税 10%）